Help Your Child

Froggy F

RICHARD & NICKY HALES
and ANDRÉ AMSTUTZ

GRANADA

Published by Dragon Books 1984
Granada Publishing Limited
8 Grafton Street, London W1X 3LA

Copyright © Richard and Nicky Hales 1984
Illustrations copyright © André Amstutz 1984

British Library Cataloguing in Publication Data

Hales, Richard
    Froggy football. – (Help your child to count; 6).
    – (A Dragon book)
    1. Numeration – Juvenile literature
    I. Title   II. Hales, Nicky   III. Series
    513'.5   QA141.3

ISBN 0 246 12467 9 (hardback)
ISBN 0 583 30728 0 (paperback)
Printed in Spain by Graficas Reunidas

Help Your Child To Start Maths

Once mathematics was the most difficult subject for children to learn. Parents would not dare do more than teach their toddlers to count to ten. Sums were for school. But maths need not be daunting; it is a part of our everyday lives. We use it in travelling, shopping and cooking. Children first meet maths in the home: in songs and rhymes, through helping in the kitchen, by playing games and just by talking about the things around them.

With your help and encouragement your children can absorb mathematical concepts while simply having fun. This series of books will help you talk about maths and suggest activities and games that will make maths child's play for you and your child.

Here are four frogs

They are called the froggy four

1 and **1** and **1** and **1** make **4**

The froggy four love playing football

I know a frog
who loves a hot dog,
fish and chips
and a chocolate log.

But what this frog
likes best of all
is to go and play
with a soccer ball.

Three frogs kick the ball,
each one wants to score.

One frog is in the goal –
three and one make four.

3 and **1** make **4**

Two frogs crash,
their heads are rather sore.

Two frogs bring a stretcher –
two and two make four.

2 and **2** make **4**

One fat frog
tying up his boot,
two fat frogs
trying hard to shoot,
one fat frog
has fallen on the floor,

one and two and one make four.

1 and **2** and **1** make **4**

One frog scores a goal,
the crowd begins to roar.

Three frogs are cheering –
one and three make four.
1 and **3** make **4**

Four fat frogs
frightening the fish,
four fat frogs
frightening the fish,
and if one fat frog
should suddenly go
    SPLISH!
There'll be
three fat frogs
frightening the fish.

Hip hip hurray!
It's Cup Final day,
and the Ratty Rangers
have come to play.

How many rats in the team?
What number will Freddy
have on his shirt?

All the frogs
from far around
have flocked into
the football ground.

Croaker scores
and time is up.
The froggy four
have won the cup!

Things To Do

Sorting
Ask your child to sort different coloured objects – buttons, counters, bricks – or a mixture of objects, into groups of a single colour. For example, ask him to make a set of blue bricks.

Card games
Play happy families, snap and other card games specially designed for children.

Dot to dot
Give your child dot to dot books, or make your own dot to dot puzzle. Let him choose a picture from this or some other book. Place some tracing paper over the picture and dot the outline. Number the dots in order and let him join them up.

COLLECTABLES

Superbikes

This is a STAR FIRE book

STAR FIRE BOOKS
Crabtree Hall, Crabtree Lane
Fulham, London SW6 6TY
United Kingdom

www.star-fire.co.uk

First published 2008

08 10 12 11 09

1 3 5 7 9 10 8 6 4 2

Star Fire is part of The Foundry Creative Media Company Limited

The CIP record for this book is available from the British Library.

ISBN: 978 1 84786 205 1

Printed in China

Thanks to: Chelsea Edwards, Andy Frostick and Nick Wells

All images courtesy of Mirco De Cet © 2008

COLLECTABLES

Superbikes

James Cadogan

STAR FIRE

Foreword

*Fast, mean and incredibly cool, superbikes
just take your breath away. Honda, Ducati,
Yamaha and Triumph amongst many other
celebrated marques have all muscled in on the
phenomenon of sheer speed and power. This little
book is an exhilarating celebration of the ultimate
sexy road machine and a feast of glorious
chrome for all enthusiasts. Just enjoy it!*

Aprilia RSV1000 Factory

Factory Location:	*Italy*
Date of manufacture:	*2007*
Engine capacity:	*997 cc*
Number of cylinders:	*V twin*
Brake horse power:	*143 bhp*
Top Speed:	*172 mph (276 kph)*

Aprilia RSV1000

Factory Location:	*Italy*
Date of manufacture:	*2002*
Engine capacity:	*998 cc*
Number of cylinders:	*V twin*
Brake horse power:	*128 bhp*
Top Speed:	*170 mph (272 kph)*

Benelli Tornado Tre 900 RS

Factory Location:	Italy
Date of manufacture:	2004
Engine capacity:	898 cc
Number of cylinders:	In line triple
Brake horse power:	133 bhp
Top Speed:	167 mph (268 kph)

Bimota Santamonica

Factory Location:	*Italy*
Date of manufacture:	*2006*
Engine capacity:	*996 cc*
Number of cylinders:	*V twin*
Brake horse power:	*133 bhp*
Top Speed:	*170 mph (273 kph)*

BMW R 1200 S

Factory Location:	*Germany*
Date of manufacture:	*2007*
Engine capacity:	*1170 cc*
Number of cylinders:	*V twin*
Brake horse power:	*122 bhp*
Top Speed:	*148 mph (238 kph)*

BMW R 1200 S

'... carved its own niche as a superb high-speed distance
bike that could carry a pillion and luggage too without
breaking a sweat. Radical Telelever front suspension's
no longer novel but still works and build quality remains
a strong point so big mileages are very possible.'

Motorcycle News

BMW K 1200 RS

Factory Location:	*Munich, Germany*
Date of manufacture:	*1998*
Engine capacity:	*1171 cc*
Number of cylinders:	*In line 4*
Brake horse power:	*130 bhp*
Top Speed:	*153 mph (245 kph)*

18

Ducati 916 SPS

Factory Location:	*Italy*
Date of manufacture:	*1998*
Engine capacity:	*996 cc*
Number of cylinders:	*V twin*
Brake horse power:	*123 bhp*
Top Speed:	*170 mph (273 kph)*

20

Ducati 999

Factory Location:	Italy
Date of manufacture:	2004
Engine capacity:	998 cc
Number of cylinders:	V twin
Brake horse power:	124 bhp
Top Speed:	165 mph (265 kph)

Ducati 748 SPS

Factory Location:	*Modena, Italy*
Date of manufacture:	*1998*
Engine capacity:	*748 cc*
Number of cylinders:	*V twin*
Brake horse power:	*103 bhp*
Top Speed:	*160 mph (256 kph)*

2004 Honda VTR SP-2

Factory Location:	Japan
Date of manufacture:	2004
Engine capacity:	999 cc
Number of cylinders:	V twin
Brake horse power:	133 bhp
Top Speed:	170 mph (273 kph)

Honda VF1000R

Factory Location:	Japan
Date of manufacture:	1985
Engine capacity:	998 cc
Number of cylinders:	4
Brake horse power:	120 bhp
Top Speed:	130 mph (209 kph)

Honda CBR1000RR FireBlade

Factory Location: Japan

Date of manufacture: 2007

Engine capacity: 998 cc

Number of cylinders: 4

Brake horse power: 170 bhp

Top Speed: 174 mph (280 kph)

30

Honda CBR1000RR FireBlade

'... delivers polished, glitch-free 180mph potency with almost no effort and yet blends this not just with the expected Honda build quality and class, but also a fat, hum-dinger, wheelie-pulling midrange that makes it more sheer fun than any since the 92 original Honda FireBlade. The latest Honda CBR1000RR FireBlade is quite simply a class act.'

Motorcycle News

Honda CBR900RR FireBlade

Factory Location:	Japan
Date of manufacture:	1997
Engine capacity:	918 cc
Number of cylinders:	4
Brake horse power:	128 bhp
Top Speed:	160 mph (257 kph)

34

Honda CBR900RR FireBlade

'The motorcycle that changed how sports motorcycles were built. The Honda CBR900RR FireBlade wiped the floor with its competitor motorcycles not by being more powerful (it wasn't) but by being lighter. Rivals were all well over 200 kg, the first Honda CBR900RR FireBlade was just 185 kg which made it perform superbly. It's still a wild ride and a surprisingly practical motorcycle too.'

Motorcycle News

2007 Honda VTR SP-2

Factory Location:	Japan
Date of manufacture:	2007
Engine capacity:	999 cc
Number of cylinders:	V twin
Brake horse power:	133 bhp
Top Speed:	170 mph (273 kph)

Honda Super Blackbird XX

Factory Location:	Japan
Date of manufacture:	2007
Engine capacity:	1137 cc
Number of cylinders:	4
Brake horse power:	164 bhp
Top Speed:	176 mph (283 kph)

Kawasaki ZX6R

Factory Location:	Japan
Date of manufacture:	2002
Engine capacity:	636 cc
Number of cylinders:	4
Brake horse power:	112 bhp
Top Speed:	162 mph (260 kph)

Kawasaki ZZR 1200

Factory Location:	Japan
Date of manufacture:	2005
Engine capacity:	1164 cc
Number of cylinders:	4
Brake horse power:	123 bhp
Top Speed:	155 mph (249 kph)

Laverda 750S Formula

Factory Location:	*Italy*
Date of manufacture:	*1999*
Engine capacity:	*747 cc*
Number of cylinders:	*Twin*
Brake horse power:	*85 bhp*
Top Speed:	*140 mph (224 kph)*

Moto Guzzi V10 Centauro

Factory Location:	*Italy*
Date of manufacture:	*1998*
Engine capacity:	*992 cc*
Number of cylinders:	*V twin*
Brake horse power:	*95 bhp*
Top Speed:	*139 mph (225 kph)*

48

MV Agusta F4 Serie Oro

Factory Location:	Italy
Date of manufacture:	1999
Engine capacity:	749.4 cc
Number of cylinders:	In line 4
Brake horse power:	126 bhp
Top Speed:	175 mph (281 kph)

2004 Suzuki GSX1300R Hayabusa

Factory Location:	*Japan*
Date of manufacture:	*2004*
Engine capacity:	*1298 cc*
Number of cylinders:	*4*
Brake horse power:	*175 bhp*
Top Speed:	*190 mph (305 kph)*

2007 Suzuki GSX1300R Hayabusa

'The Suzuki Hayabusa has possibly the most powerful
production engine on the market, alongside the Kawasaki
ZZ-R1400, Suzuki are quoting 194 bhp and 115 ft lb of
torque for the Hayabusa. The motorcycle isn't restricted
in the lower gears, it's full power all the way to the
186 mph speed limiter.'

Motorcycle News

2007 Suzuki GSX1400

Factory Location:	*Japan*
Date of manufacture:	*2007*
Engine capacity:	*1402 cc*
Number of cylinders:	*4*
Brake horse power:	*105 bhp*
Top Speed:	*145 mph (233 kph)*

2007 Suzuki GSX-R1000

Factory Location: Japan

Date of manufacture: 2007

Engine capacity: 999 cc

Number of cylinders: 4

Brake horse power: 185 bhp

Top Speed: 190 mph (305 kph)

2007 Suzuki GSX-R1000

'Suzuki has never made a bad GSX-R1000, and they still haven't. Their flagship superbike always leads the class, never follows, and the new GSX-R1000K7 is no different. Not only does it have more power than ever before – up 7 bhp to 185 bhp – it's now even to handle than ever.'

Motorcycle News

2004 Triumph Daytona 955i

Factory Location:	*UK*
Date of manufacture:	*2004*
Engine capacity:	*955 cc*
Number of cylinders:	*In line triple*
Brake horse power:	*147 bhp*
Top Speed:	*165 mph (265 kph)*

1997 Triumph Daytona T595

Factory Location:	UK
Date of manufacture:	1997
Engine capacity:	955 cc
Number of cylinders:	In line triple
Brake horse power:	125 bhp
Top Speed:	161 mph (258 kph)

Triumph Rocket III

Factory Location:	UK
Date of manufacture:	2007
Engine capacity:	2294 cc
Number of cylinders:	In line triple
Brake horse power:	120 bhp
Top Speed:	135 mph (217 kph)

Triumph Rocket III

'The Triumph Rocket III is the biggest, most bad-ass motorcycle money can buy. The specs are awesome a 2.3-litre engine producing almost 150 ft lb of torque, pistons the same size as those found in a Dodge Viper supercar and what was at one time the biggest back tyre on a production bike… The Triumph Rocket III is simply an incredible experience and bravo to Triumph for making it.'

Motorcycle News

Yamaha YZF-R1

Factory Location:	*Japan*
Date of manufacture:	*2002*
Engine capacity:	*998 cc*
Number of cylinders:	*4*
Brake horse power:	*150 bhp*
Top Speed:	*175 mph (281 kph)*